INGENIOUS GENES

MICROEXPLORERS

First edition for the United States and Canada
published exlusively 1997 by Barron's Educational Series, Inc.

Originally published in English under the title *Ingenious Genes*
© Copyright USEFUL BOOKS, S.L., 1997 Barcelona, Spain.

Authors: Patrick A. Baeuerle and Norbert Landa
Illustrators: Antonio Muñoz and Roser Rius
Graphic Design: IGS – Barcelona, Spain

Address all inquiries to:
Barron's Educational Series, Inc.
250 Wireless Boulevard
Hauppauge, NY 11788

Library of Congress Catalog Card No. 97-74825

International Standard Book Number 0-7641-5063-4

Printed in Spain

987654321

INGENIOUS GENES

MICROEXPLORERS

Learning about the
fantastic skills
of genetic engineers
and watching them
at work

Patrick A. Baeuerle
and Norbert Landa

Welcome to the

Has anyone here never cut his or her skin? No?

We all know about cutting ourselves. A cut hurts, and blood drips out—but only for a short time. Very soon it clots and becomes solid. The blood dresses its own wound. Being able to change from a liquid to a solid mass at the right time and place is one of blood's most fascinating properties.

Unfortunately, some people have blood that doesn't do this. They suffer from a disease called hemophilia.

What is the difference between the blood of hemophiliacs and the blood of healthy people? Hemophiliacs lack something essential—a protein that controls blood clotting. It isn't in their blood because of a very tiny change in one of their genes.

Our genes tell the cells of our body what proteins to make. Genes are unbelievably tiny threads contained in each of our cells. For example, genes tell the cells to make special proteins that let blood clot at certain times. Faulty genes cannot do this. However, geneticists today are seeking ways to cure hemophilia and many other diseases caused by faulty genes.

Let's take a closer look at genes. We must shrink very small by using our MicroMachine. First, we will encounter face-to-face what blood consists of and see how it can turn into a solid mass. After shrinking further, we will watch genes at work.

Scientists have learned that genes work the same way within humans, dogs, carrots, and

tour!

even bacteria. So genetic engineers can cut, copy, and paste genes among bacteria, plants, and animals. They can handle things that are about one-millionth of a yard long and a billionth of a yard wide. Imagine, they can paste human genes into bacteria and tobacco plants and let them produce medicine for human patients!

On our expedition, we will learn how geneticists do all this and more. We will learn what transferring genes in the laboratory has in common with breeding animals and plants. We will learn even more about the unique properties of human genes—why each of us has his or her very own genetic fingerprint and how this can solve puzzling crimes and historical mysteries.

By the way, I am your tour guide. Just call me Gene. Do not hesitate to ask me questions during our expedition.

Shhhhrink!

Ouch!!!

A small cut in the skin hurts. Pain is the body's way of warning us to take good care of ourselves, especially of our blood. We need blood badly; it has many essential jobs to do.

Blood contains many parts. Here we see red blood cells. There are more of them in each of us than people on our planet. Red blood cells pick up oxygen that comes into our lungs with the air we breathe. Red blood cells deliver the oxygen to other cells all around the body.

These pale things are white blood cells. It is their job to destroy alien intruders such as bacteria that might try to thrive in our body.

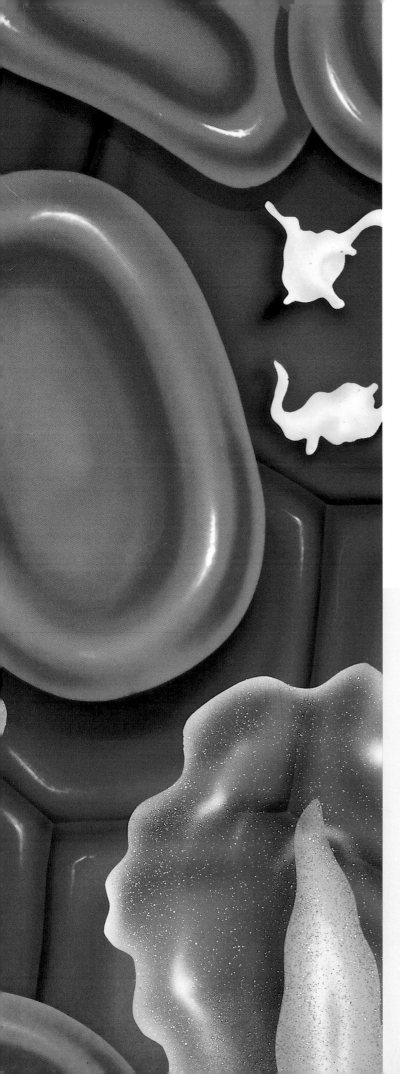

These strange looking cells are thrombocytes. We have millions of them in each drop of blood. They are responsible for making blood become solid near a wound. These cells—and many more things—constantly drift through a watery liquid called blood plasma. Blood is a liquid organ and is just as important as other organs like the heart or the lungs.

Why is blood liquid, Gene?

Liquids can be pumped, but solid matter cannot. Blood has to carry things to cells in all parts of the body, even to the tips of our toes. Blood must flow in a tube system called blood vessels. Once a vessel becomes damaged, liquid blood could leak out. Losing blood is dangerous. In case of emergency, blood needs to become solid quickly at the site of injury.

What makes blood become solid?

See the thrombocytes? They do three things for us. First, they stick to a leaky site in a blood vessel and fill the gap loosely. Then they release substances that make the blood vessel contract. Now the blood flows more slowly. Finally, thrombocytes and long, sticky protein fibers called fibrin form a meshwork that closes the gap in the blood vessels tightly. Where do those fibers come from?

Fibrin fibers

We never know when and where leaks in the blood vessels will need to be fixed. Blood must always carry along the substances that could quickly form sticky fibers and let blood solidify in an emergency. Therefore, these substances are proteins called fibrinogens or *fibrin makers*. Only during an injury is a signal sent out that puts a dozen different kinds of proteins to work. Those long and sticky fibrin fibers soon appear at the right place as a result of binding the fibrin proteins together.

If just one out of a dozen different control proteins in the series isn't made properly or is missing, fibrin fibers cannot be formed. As a result, the loosely fixed hole in the blood vessel will never get sealed by fibrin fibers. The gap between the blood vessel cells will not be closed. The injured person will continue to lose blood. This is what happens with hemophiliacs.

Can a hemophiliac die from a wound?

This can happen only if a doctor isn't around to help. He or she can give the bleeding person the missing protein, which has been isolated from thousands of pints of blood donated by healthy people. Isolating the protein, however, requires a tremendous effort, and the treatment costs a lot of money.

Can blood also clot in our blood vessels without an injury?

Oh, yes. However, this would plug the blood vessels and stop the blood from flowing. In fact, that is what happens to patients during a stroke. It is good that blood clotting is so complicated. Clotting should only happen at the right site in the right moment and not by accident at the wrong places. Now, let's have a closer look at genes. We'll see how they tell our cells what to do and why they can be so important for our well-being.

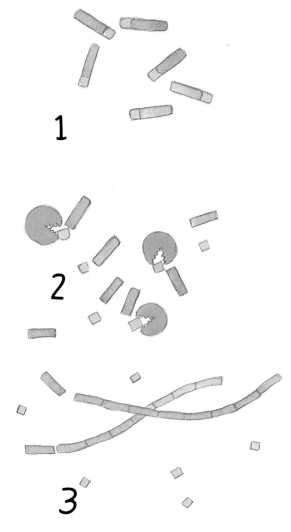

This is how fibrin fibers are made:
1) *In liquid blood, there are proteins called fibrinogen drifting around.*
2) *Proteins controlling blood clotting appear at the site of an injury. One of them chops off tiny bits from the fibrin proteins.*
3) *The cut fibrin clings together to form sticky fibrin fibers that make the blood clot.*

Clever cells

Our bodies, along with those of every creature living on earth, are formed entirely from cells. Cells are so small that it would take more than 2,500 of them to cover an inch. Each cell is alive. Together, many form an organism. Humans have red and white blood cells, nerve cells, muscle cells, skin cells, and hundreds of other kinds of cells of varying shapes that perform various tasks. Trillions of cells form every hard and soft part of your body. Some move through your blood, some make you think and feel, and others contract so you can move.

Cells, the building blocks of our body, can do amazing things. Cells produce most everything they consist of. One of the essential things they need—and make—are called proteins. Fibrin is just one of the 100,000 kinds of proteins made in human cells. Here we can look into a white blood cell. The cover, called a membrane, has been partly removed so we can see the cell's inner parts. From the inside, nearly all kinds of cells look quite the same.

What do proteins do, Gene?

Some are needed to form a cell and all its parts. Some act as couriers for exchanging messages between cells. The proteins called enzymes initiate the many thousands of reactions that keep a cell alive. A healthy body depends on the continuous interplay of about 100,000 different proteins in just the right amounts and in just the right places to do just the right jobs. Just think about fibrin forming fibers to close a leaky blood vessel.

We humans consist of countless cells, which is something we have in common with other beings like apple trees and dogs. However, some organisms, which we can't see with the naked eye, are as small as or even smaller than our body cells. They are bacteria. Bacteria are single-celled organisms. Many kinds are harmless or even useful. Some of them live on our skin, in our mouth, or in our intestines. Other kinds are dangerous and can cause diseases.

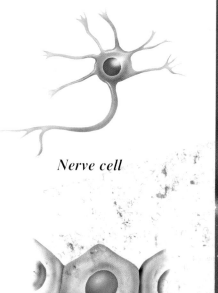

Nerve cell

Skin cell

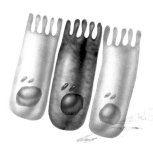

Intestine cell

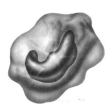

White blood cell

Bacteria

Bacteria consist of only one single cell that eats, multiplies, and makes proteins. What bacteria do and how they do it is quite similar to what our body cells do—except that bacteria do it for their own sake. In contrast, our cells work together for the sake of the whole organism.

All cells—bacteria, liver cells, skin cells, and white blood cells—have tiny parts called ribosomes that work as protein-producing factories. They assemble substances called amino acids in specific sequence to form specific proteins.

They do this according to the protein recipes stored on genes within the nucleus of each cell.

Bacteria and human body cells make different proteins because they have different genes. This is why bacteria look and behave like bacteria and why humans look and behave like humans. Bacteria, for example, cannot make fibrin. On the other hand, our cells cannot make the very special enzymes that help some bacteria eat mineral oil. Bacteria have no fibrin gene, and we don't have the genes for such unusual enzymes.

Ingenious *genes*

Without genes a cell could live only for a short time. A cell needs instructions telling it what to do—which proteins to make. Otherwise, it would soon run out of fresh proteins and die.

Genes are like sentences written in a chemical language and lined up on a thread called DNA (deoxyribonucleic acid). In each human cell, the DNA is split into 46 pieces called chromosomes, which are wrapped up and stored in the cell's nucleus. Every human cell contains the same 100,000 genes, collectively called the genome. Each gene can tell the cell precisely how to make a particular protein. You can imagine that a liver cell's job is different from that of a skin cell or a white blood cell. So liver cells switch on a different set of genes than do skin cells and, therefore, make other proteins with other functions. However, many genes are common to both skin and liver cells and even to bacteria. These *housekeeping* genes make proteins that every cell needs for its basic functions such as cell division or energy production.

Where do cells get their genes from, Gene?

All human cells develop from just one fertilized egg cell, which is in the mother's womb. At first, an unfertilized egg cell contains only half the genes it would need to grow and develop. The other half comes with the father's sperm cell. When both cells meet and fuse, the egg cell is fertilized. It contains a unique and complete set of genes. Immediately, the egg cell starts to divide and to build a tiny human being with all the different kinds of cells a body needs. When a cell divides, it makes a precise copy of its entire DNA. All the genes are written in the DNA thread. In this way, all of our cells have the same set of genes. In fact, we have two copies of each gene—one from the mother's egg cell and one from the father's sperm cell.

are inside a cell now. The big blue ball is the cell's
cleus, which contains and safeguards the chromo-
nes. Here you see how a thread of DNA would look if it
re pulled out of the nucleus. Each rung of the DNA
der consists of two out of four possible nucleotides—
h shown in a different color.

This single thread is called messenger
RNA. It carries the true copy of a gene—a
recipe for making a particular protein. It
takes the recipe out of the nucleus and
brings it to the protein factories of the cell.

The genetic *code*

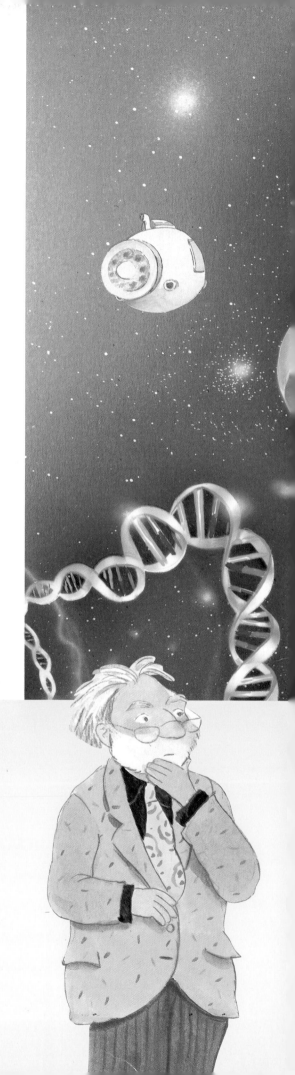

Altogether, the DNA thread stored in the nucleus of a human cell contains three billion letters. Most of the DNA thread does not store gene recipes but, instead, contains nonsense. Separated a large distance from each other are meaningful parts—the genes.

The interesting parts of the twisted DNA ladder are the steps—the DNA letters. They are made of four quite similar chemical substances called nucleotides, nicknamed A, T, G, and C. Each step consists either of an A linked to a T or a C linked to a G. The nucleotides have a strict buddy system—A, for instance, never links with C or G, and C never with T or A. If this ladder splits in the middle—which happens before cells divide—there is only one way to fix the two halves and make two identical copies of the ladder. This is the way cells pass on their genes to their daugther cells. This is also how gene copies are made and stored on messenger RNA threads that bring protein recipes to the ribosomes.

A gene is nothing but a piece of the DNA thread containing the recipe for how to make a particular protein. The nucleotide language of the recipe consists only of four chemicals—A, T, G, and C. Proteins, however, consist of 20 different kinds of building blocks called amino acids. They must be put together in the right sequence and quantity to form a particular protein. How can just four different DNA letters describe 20 different amino acids? There must be a trick!

What's the trick, Gene?

The trick is to combine three nucleotide letters to describe one amino acid. This is the fabulous genetic code—a group of three DNA letters makes one DNA word that stands for one particular amino acid. The sequence of these DNA words clearly tells the cell in which sequence to put together the amino acids that build a particular protein.

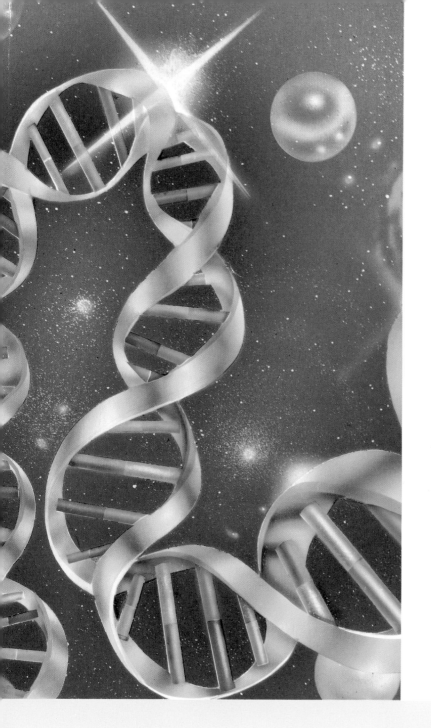

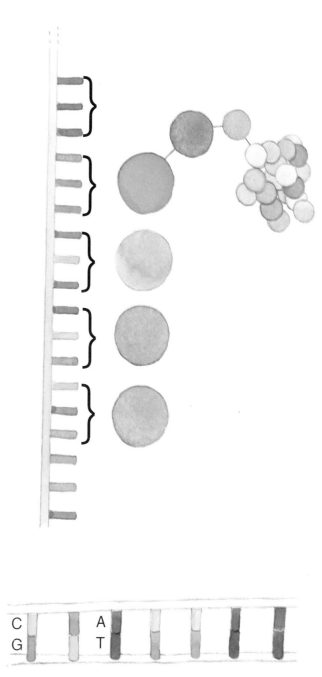

Altogether, four different nucleotide letters can form 64 possible words, for instance AAA, AAT, TAA, ATA, ATT, TTA, ATG, CGA, TTT, and so on. Since the genetic code needs to describe only 20 amino acids, some of these three-letter nucleotide words describe the same amino acids. Other nucleotide words signal the beginning of a gene or where the gene stops. Other signals tell the proteins that regulate genes when to turn on the genes—when to make a copy of the gene and bring it to the protein factories.

Do only humans use the genetic code?

No, the genetic code is universal. One of the fascinating facts about the genetic code is that all creatures on planet Earth use it. All living beings we know of, no matter if they are bacteria, apple trees, dogs, or humans, use the same genetic code for the same 20 amino acids.

Some genes that stand for housekeeping proteins even have very similar sequences in humans and bacteria. We have quite a lot in common with all the creatures on our planet!

How to make
proteins

The genes tell the ribosomes to pick certain kinds of amino acids and put them next to other kinds in order to make a specific protein the body needs. First, a copy of the gene, called messenger RNA, is made. It travels out of the nucleus and gives its instructions to the ribosomes. The ribosomes can read the genetic code. They catch the amino acids drifting in the cell and align them in the right sequence and in the right number—like beads on a string. In order to make a particular protein, a ribosome must assemble the amino acids in a particular sequence—according to the recipe provided by a particular gene.

Can proteins be made only by cells?

Yes. All proteins consist of building blocks called amino acids. There are 20 different kinds of amino acids. Only a cell's ribosomes put them together to form proteins.

Proteins are incredibly tiny particles. Line up one million proteins and they would cover less than one-tenth of an inch! Despite their small size, proteins are quite complicated things—so complicated that only cells can build them. All proteins of a particular kind look and behave exactly the same. Some kinds look like tiny balls, others like rods or fibers, and some like tubes.

The freshly made protein thread instantly starts folding itself into its final shape, which is the key to its function. Now the fresh protein needs to find its proper place in the cell. The protein fibrin, for instance, must be transported out of the cell in which it was made to do its job in the blood. Frequently, proteins need to find other proteins of either the same or a different kind so they can do their jobs jointly.

Some proteins consist of only dozens and others of thousands of amino acids. There are 100,000 different kinds of proteins at work within a human body. How big proteins are, what they look like, and what they can do for the body depends on the sequence in which the 20 different amino acids are lined up.

Faulty genes, *faulty* proteins

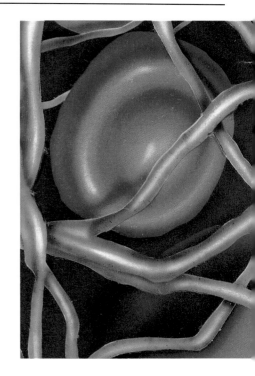

When billions of fresh cells are made from a single one and billions of genes are copied each time, some of the genetic words will likely be misspelled. A cell carrying a gene with a spelling error may not make the right protein, just as a telephone number won't make the right connection if you dial even a single wrong digit.

Most of the time, such a genetic mistake doesn't matter because countless other cells are making the right protein. The cell with the faulty gene eventually dies and does not pass on its defect. However, if the defect occurs in genes that control how often a cell divides, these faulty genes can suddenly cause the cell to divide much faster than all the other cells. A cell clump may develop that doctors call a tumor. A big problem arises if a gene in the egg cell or the sperm cell is faulty. This faulty gene will be given to all the cells of the tiny growing organism.

Remember, we have two copies of each gene, one from the mother and one from the father. The worst case happens when both copies the child gets from mother and father are faulty. If the child doesn't receive a healthy gene to produce a proper protein, the child could have a genetic disease like hemophilia.

A child can have a genetic disease the parents don't have. This happens when each parent has both a faulty gene and a healthy gene. Most parents do not even know that they have a faulty gene because their healthy gene still makes the proper protein.

Does each faulty gene cause a disease?

Not at all! Sometimes a spelling error (also called a mutation) can make a protein function even better. This will make the entire living being work better. A person with a mutation may run faster than others in the thin air of a high altitude. She or he might be better protected against a particular disease such as malaria. Whatever the advantage is, it will be passed on to the person's children, grandchildren, and great-grandchildren.

Genes are recipes for making all the proteins a cell needs to perform its functions. Proteins make a big and complicated organism like you and me. They keep us alive.

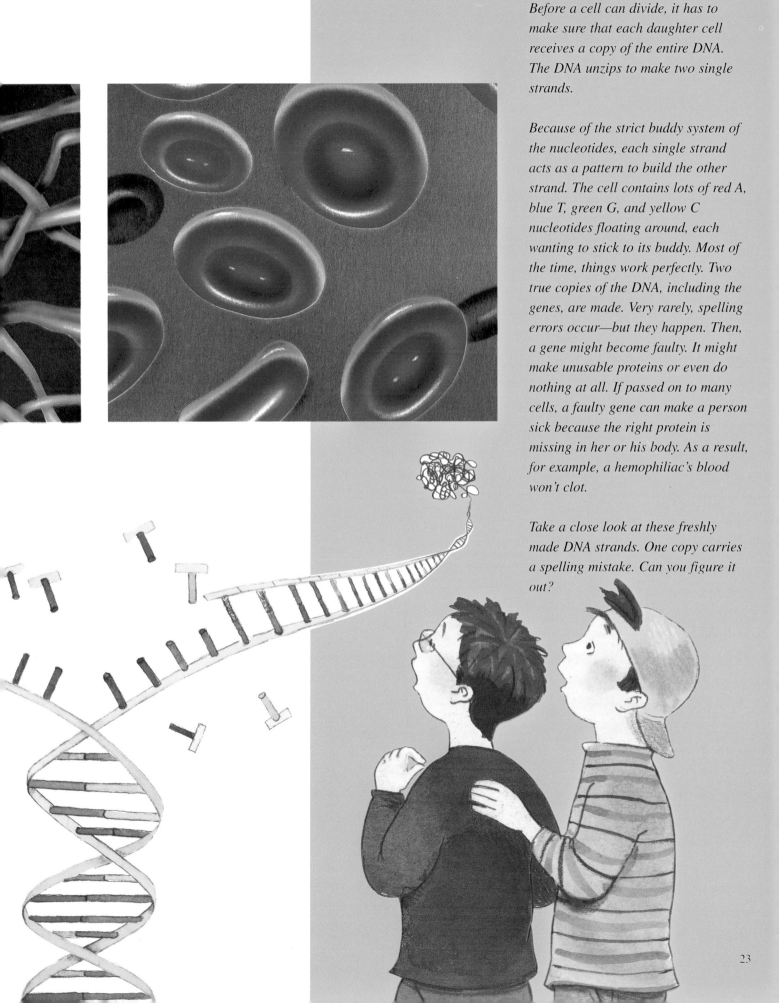

Before a cell can divide, it has to make sure that each daughter cell receives a copy of the entire DNA. The DNA unzips to make two single strands.

Because of the strict buddy system of the nucleotides, each single strand acts as a pattern to build the other strand. The cell contains lots of red A, blue T, green G, and yellow C nucleotides floating around, each wanting to stick to its buddy. Most of the time, things work perfectly. Two true copies of the DNA, including the genes, are made. Very rarely, spelling errors occur—but they happen. Then, a gene might become faulty. It might make unusable proteins or even do nothing at all. If passed on to many cells, a faulty gene can make a person sick because the right protein is missing in her or his body. As a result, for example, a hemophiliac's blood won't clot.

Take a close look at these freshly made DNA strands. One copy carries a spelling mistake. Can you figure it out?

23

Gene and protein
medicine

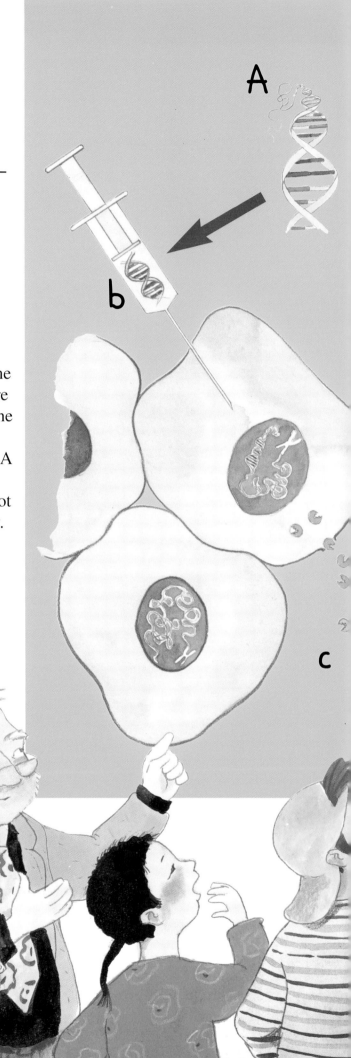

Let's see how we can use this knowledge about genes and proteins to help these poor kids whose blood doesn't stop flowing through a wound—and also many other ill people. Faulty genes cause a few hundred hereditary diseases, and hemophilia is just one of them.

Can't we repair a faulty gene?

Unfortunately, no. Even if we knew exactly which one of the many thousand of DNA letters in the gene were misspelled, it would be too complicated to take out the wrong nucleotide from the DNA and put in the right one. In fact, we can't even see a cell's nucleus or DNA with the naked eye. Besides, it would be useless to repair a faulty gene in only a few cells. They would not make enough of the protein to supply the entire body.

Why not give cells healthy copies of genes?

That's a good idea. Actually, genetic engineers are trying practical ways to do that. For example, they use tiny needles to inject DNA carrying the healthy gene into cells so the cells could produce the right protein. However, there is one problem. One cell alone doesn't have the faulty gene. All of them do—billions of cells. Treating each cell separately is not practical. Instead, geneticists would need millions of tiny ferries to bring millions of good genes into millions of cells, hoping enough of the right protein is produced for a sufficiently long time. There are such genetic ferries, as we'll see

B

C

Geneticists can help a person that lacks blood-clotting proteins.

(a) This is a healthy human blood-clotting gene taken out of any cell of a healthy person.

(b) In the lab, the healthy gene is placed into bacteria. Here the human gene is built into the bacterial DNA. The bacteria now produce human blood-clotting proteins.

(c) Doctors inject the protein medicine into a sick person.

(d) Alternatively, the healthy gene is transferred directly into the body cells of the person with faulty genes.

(e) The healthy gene is built into the DNA of the sick person and will replace the faulty gene. Now the body cell can make blood-clotting proteins. This treatment is called gene medicine.

(f) In case of a wound, the proteins can make the blood clot.

later. For now, what else could scientists do to provide hemophiliac patients with the blood-clotting protein they badly need?

Can scientists take the protein from other people or from animals?

Yes, they have been doing that for many years. The right blood-clotting proteins can be filtered out of the blood of human donors and given to patients. Sometimes, this can be quite dangerous because donors can carry human viruses that cause deadly diseases. The viruses can still be in an isolated protein given to hemophiliacs. Here is another hint. All cells use the same genetic code and the same amino acids. Think of what foreign cells would do when given the human gene.

Can other beings produce the human protein?

Wow! What a clever idea! A hemophiliac patient's own cells cannot produce a particular protein because the genes are faulty. The cells of other beings also cannot do it simply because they do not have the particular human gene. Genetic engineers can take a healthy gene from a human cell and put it into a foreign cell. This cell will divide over and over again, copying the human gene many times. Genetic engineers give the cells the right signals to turn on the gene. Then these foreign cells with the human gene will produce, along with their own proteins, masses of the human blood-clotting protein. Finally, genetic engineers make a lot of the protein and give it to the patient's blood.

What a wonderful idea—and it works! This is how genetic engineers alter the genes of plant, animal, and bacterial cells to make them produce human protein medicines.

Of humans and *sheep*

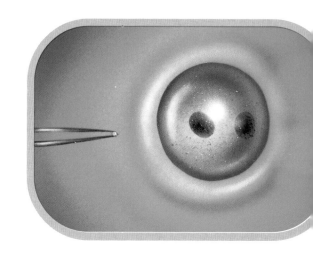

Genetic engineering works because all cells existing on this planet use the same language for their genes and proteins. They all are, in a way, tiny protein factories doing what their genes tell them. Different genes make different proteins that make different cells that make different beings. A sheep's mammary gland cell in its udder, for example, would not make a human blood-clotting protein, not unless it were told to do so by an additional human gene. This is how genetic engineering works.

Geneticists know exactly which human gene is responsible for making which blood-clotting protein. They can find and cut out the healthy gene from human DNA. Then they give it all the signals the gene needs to be turned on only in a particular kind of cell—in sheep mammary gland cells that make all the proteins found in sheep's milk.

Geneticists, using an incredibly small needle, place the human gene with all its DNA signals into a sheep's fertilized egg cell. There the tiny piece of human DNA will be built into the sheep's DNA. This changes the sheep's genome. From now on it contains, in addition to all the recipes for sheep proteins, the recipe for this particular human protein. The egg cell in the mother sheep divides over and over again. The human gene is in all the daughter cells. A couple of months later, the mother sheep will give birth to a lamb that looks, feels, and behaves normally.

When the lamb has grown up and becomes a mother sheep, the milk it gives is a bit different from the milk other sheep give. It contains not only the normal sheep milk proteins but also the human blood-clotting protein. Sheep do not need this protein, but the hemophiliac with the badly cut finger does.

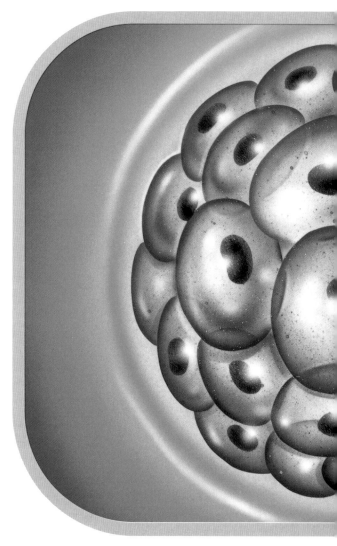

The human protein made in the sheep's udder was manufactured according to instructions given by the human gene. Therefore, the protein looks and works like it was made in a human body. The blood-clotting protein can be filtered out of the milk and injected into the blood of the patient. Now the patient's blood can clot just like the blood of healthy people.

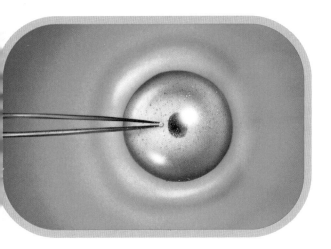

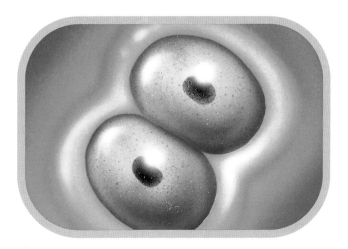

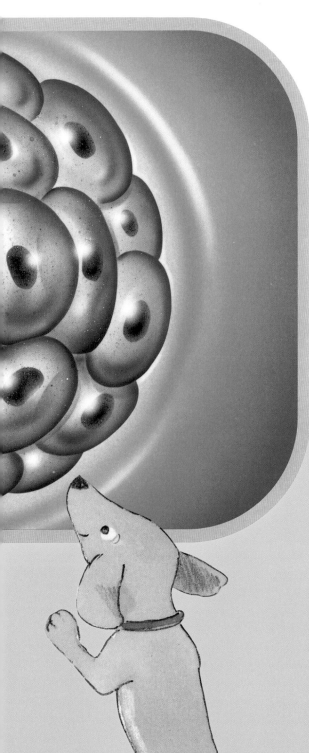

What happens to the lambs of this sheep?

The lambs will inherit the human gene from their mother, just like any other sheep gene. We call such an animal transgenic. It carries genes that originally came from other species but now permanently belongs to it—and to its offspring. A transgenic sheep can pass on all its genes, including the human one, to its lambs in just the same way as those genes that make a sheep bleat instead of bark.

Could geneticists make sheep that bark, Gene?

Probably not. The ability to bleat or bark depends on hundreds of genes and their proteins. They act together in a way much too complicated to figure out. Besides, what would be the point in having a sheep that barks? Confusing the shepherd's dog?

Are genes transferred in nature as well?

Yes. Many viruses, after entering a body cell, will insert their genes into the host's DNA. Sometimes, this can cause the cell to divide endlessly and grow into a tumor. In other cases, viruses just hide and sleep in host cells. Some viruses are even inherited by offspring in this way, making the animal naturally transgenic. In contrast with the messy and sometimes dangerous way viruses insert their genes, genetic engineers know where and how to build genes into the DNA of a transgenic animal. They do not want their precious animals to get sick, do they?

27

Of humans, bacteria, and *tobacco* plants

In many diseases caused by missing proteins, many genes are involved. This makes it very difficult to find a simple cure by just supplying the healthy protein or gene. Doctors have to do something else.

Diabetes is one disease where not enough protein is present. Diabetes is not caused by a faulty gene but by damage to the cells producing a protein called insulin. Normally, insulin is made by special cells of the pancreas, a gland in the abdomen. The hormone insulin tells the liver to take extra sugar out of the blood and store it as a starch. If insulin is missing, most of the sugar stays in the blood. This can seriously damage blood vessels, eyes, and organs. Diabetes patients are allowed to eat only very few sweet things. They must inject insulin into their blood on a daily basis to keep their blood sugar level normal.

Where does the injected insulin comes from, Gene?

Until recently, it came from the slaughterhouse! Insulin for patients used to be extracted from the pancreas of pigs and cows. Nowadays, human insulin is harvested from genetically altered bacteria that are grown and kept in large vessels. Genetic engineers have implanted into bacteria the gene that, in humans, makes the pancreas cells produce insulin. Now the bacteria are busy making human insulin!

What happens if these bacteria escape?

Most likely, genetically altered bacteria won't survive outside the lab. They would have to compete with the wild bacteria for food. Wild bacteria are not forced by their genes to make large amounts of a protein they have absolutely no use for. So the wild ones would soon outdo the altered bacteria, just like wolves would outdo a runaway pack of poodles or short-legged dachshunds in the wilderness. Moreover, even surviving bacteria with altered genes would cause no problems. It is hard to tell how they

A human gene (a) can be put into a plant cell (b), a bacterium (c), or even any body cell, such as a skin cell (d). These cells start producing the protein (e) needed.

d

e

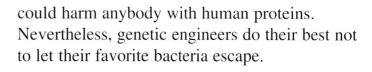

could harm anybody with human proteins. Nevertheless, genetic engineers do their best not to let their favorite bacteria escape.

Can plants read human genes, too?

Of course they understand the genetic code. For example, genetic engineers introduce genes into tobacco plants that tell the plants how to make human antibodies. Antibodies are proteins that our immune cells need to fight bacteria and viruses that invade the body. Antibodies can also bind to and remove dangerous substances from the blood. We use antibodies as medicine. Farmers crop tobacco plants that have the particular human antibody genes. The tobacco

c

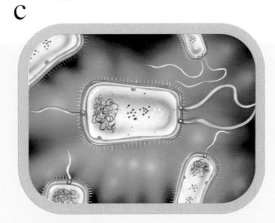

plants make human antibodies that can be filtered out and injected into patients. At last, there is a healthy use for tobacco!

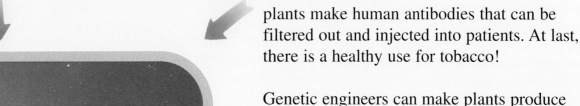

Genetic engineers can make plants produce other useful things too, such as the plants' very own pesticides. Sugar beets, for example, are constantly at risk of being eaten by a particular worm. Other kinds of beets produce a protein that makes them resistant to this very pest. So you can easily figure out what geneticists do. They take the particular gene from the resistant beet and put it into sugar beet—thus making sugar beets resistant to the nasty worm.

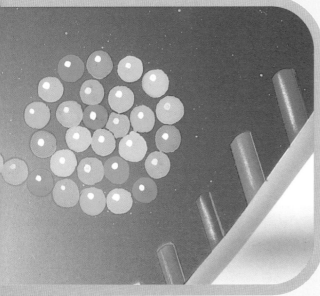

Viruses
as medicine?

Genetically changed animal cells, plant cells, and bacteria can produce proteins that people with certain diseases cannot make on their own. Such treatment helps these people to live a better life, but it doesn't really cure them of the disease. To cure them, a kind of shuttle or ferry would have to put the proper healthy gene into millions of cells with faulty genes. Viruses would be ideal gene shuttles.

Viruses make us ill, don't they?

Normally, yes. A virus is a tiny, tiny particle consisting of a protein shell with its own viral genes inside. Its only goal is to invade a body cell and introduce the viral genes. The infected cell starts to make viral proteins, forced to by the viral genes. There are countless kinds of viruses. Some cause the flu, measles, or really serious diseases such as rabies. However, geneticists know how to change the genes of a virus so that it would make ill people healthy instead of making healthy people ill.

How can a virus be tamed, Gene?

First of all, geneticists must learn the function of each viral gene and understand how the virus causes a disease. Then, they remove the very genes from the virus that make it dangerous. At the same time, they leave enough genes so that the virus can still work as a shuttle for healthy human genes. This way, geneticists convert a nasty virus to a useful servant.

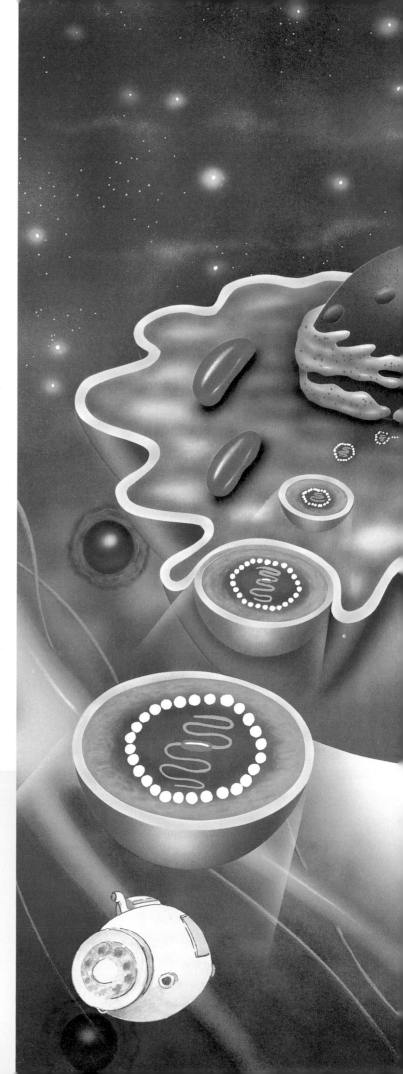

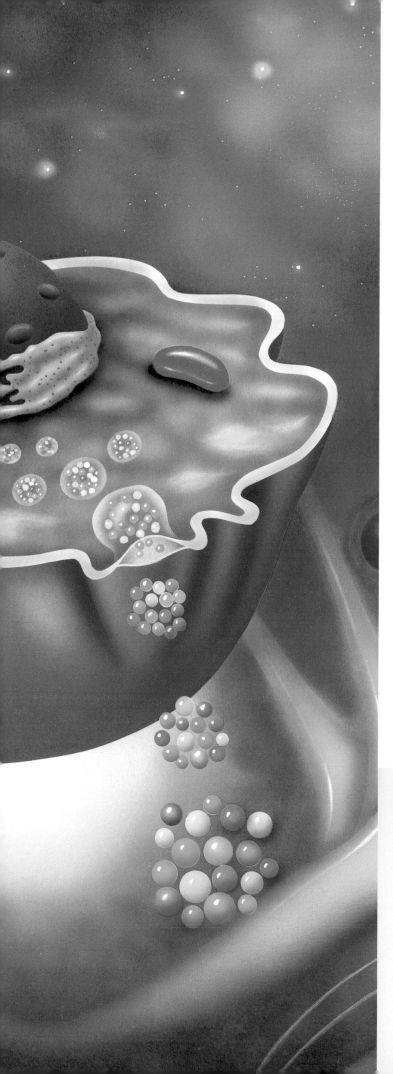

Here we can see how a virus brings its genes, including a healthy human gene, into a sick human body cell. Now the cell can start to produce the right proteins, which soon will drift out of the cell to do their job. Normally, both the cell and the virus are covered by membranes. Here part of the membranes have been removed so we can better see what happens.

The healthy human gene is then added to the now harmless viral genes. In cells kept outside the body, large amounts of the virus shuttles are produced. Doctors collect and inject these shuttles into a sick person. Like wild viruses, the shuttle viruses now start to infect cells. They do not overwhelm the cells and force them to make new viruses. Instead, the shuttle viruses bring the healthy human genes.

If everything goes according to plan, the infected cells start to make the right proteins. The patient feels better.

Have people ever been cured this way?

Since this is a novel and very complicated technique, it hasn't worked any miracles so far. Few cells get infected. The healthy protein is made in amounts too low and for too short a period of time. Many more years are needed to perfect this technique. Imagine—this way sick body cells with faulty genes could learn how to make the right proteins themselves. They would obey the healthy genes brought in by good viruses.

Will the treated patient have healthy children?

Unfortunately, no. Only a few body cells will receive the healthy gene. In order to pass it on to the offspring, the healthy gene must get into egg or sperm cells as well. Experiments to test whether this is possible are not allowed with humans. Things could go wrong and make sick children.

Cutting and pasting *genes*

Genetic engineering sounds quite simple—take a gene out here and put it in there. In reality, this is an incredibly complicated procedure. A DNA thread containing genes is so thin that one million threads laid down side by side would cover less than one-tenth of an inch. Also, DNA contains billions of genetic letters. It is like a huge library stuffed with thousands of books—all in the nucleus of a cell. So geneticists need to have very special tools and techniques to sort pieces of DNA and find the right genes.

What do these tools look like?

DNA is cared for by proteins that constantly maintain the genetic library within the cell nucleus. They unwind and rewind the DNA helix, they mend broken parts, they repair mismatches in the ladder, they tear apart the DNA ladder, they copy all the genes or just a single one, and they do many more complicated jobs.

Genetic engineers make use of these proteins, especially of one particular kind called restriction enzymes. These proteins work like scissors. For example, an enzyme named Eco R1 specializes in cutting exactly between the G and A in a DNA sequence that reads GAATTC. Other restriction enzymes cut the DNA at other sites. Geneticists use these tiny tools to cut out particular parts of the DNA. Take some DNA, add some Eco R1, and the enzyme will happily drift along the DNA until it reads GAATTC and then—chop!

How can geneticists paste genes?

The pieces of DNA cut out by Eco R1 now can be used for further experiments. For instance, let's take the piece containing the human gene for insulin. We want to paste a piece of the insulin gene into the

a) Take a DNA plasmid out of bacterium.
b) Cut open the plasmid and add a human gene.
c) The bacterial DNA plasmid now also contains a human gene.
d) Put the plasmid back into a bacterium.
e) Let the bacterium multiply, harvest the human protein, and give it to sick people.
f) Instead, do the same thing with many different human genes and create a plasmid library containing human genes stored on bacterial plasmids.

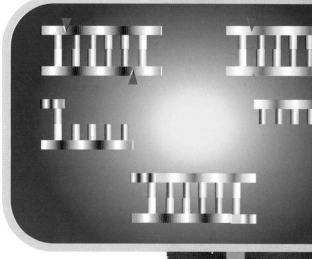

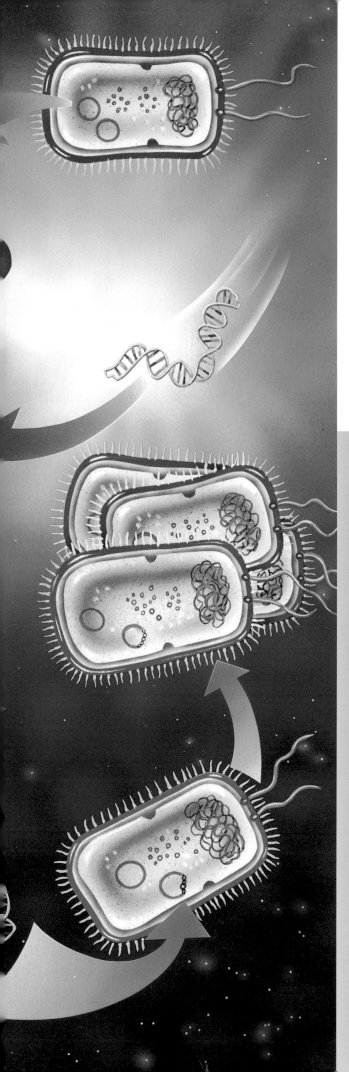

DNA of a bacterium. Eco R1 doesn't cut DNA bluntly. On each side of the cut, it leaves four nucleotides that aren't attached to a buddy. These are called sticky ends. Genetic engineers paste the sticky ends from bacterial and human DNA together.

Genetic engineers need many specific DNA pieces for all this cutting and pasting. Therefore, the pieces of DNA from a human cell must first be produced in large quantities. This job is done by bacteria.

Bacteria store a small part of their DNA in the form of a small circle called plasmid DNA—very much to the delight of the genetic engineers. Scientists can easily take plasmid DNA out of bacteria without bothering the bulk of the bacterial DNA.

Let's remove a plasmid DNA from a bacterium and cut it open, for instance with Eco R1. The plasmid becomes linear. We add the human insulin gene and some ligase, an enzyme that can paste DNA. We get back circular plasmid DNA that is a bit larger than before because it also contains the human gene. Let's put the plasmid back into the bacteria. When we feed the bacterium with a broth, it will divide over and over again, soon making billions of bacteria. They all have a plasmid with the human insulin gene. We can add some signals to tell the bacteria to turn on the insulin gene in the plasmid, and now the bacteria will produce insulin. It is also quite easy to take the plasmid DNA out of all the freshly made bacteria and add other useful pieces, for instance, those that make the gene work with a virus.

DNA that contains pieces from different species, such as humans and bacteria, is called recombinant DNA. It works a bit like a recording patched together from different tapes—with songs that come from different musicians. The songs are the genes, and the cassette player is a cell. Of course, a cell doesn't play music—it makes proteins!

How scientists make
clones

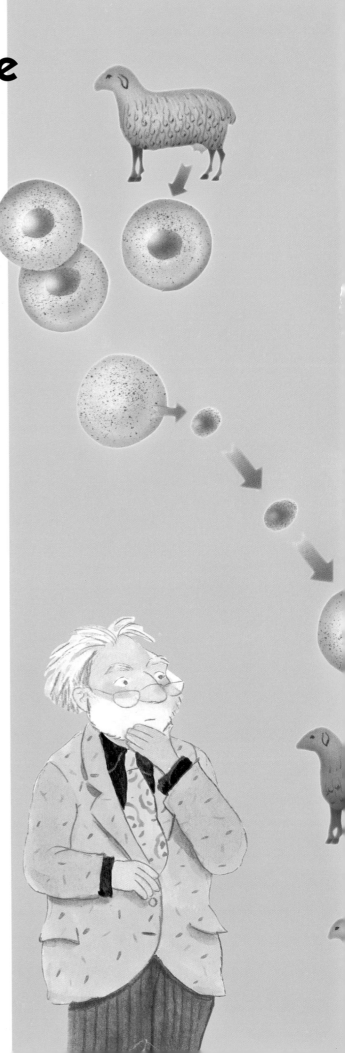

Clones are living beings that have exactly the same set of genes. In a way, almost all our body cells are clones since they all have the same genes.

Various body cells look and work differently because different genes are turned on or off. That is why our liver cells are different from our skin cells. Bacteria also are clones. The offspring have the same genes as the parent cell. Usually, cell division produces clones.

Are there also human clones, Gene?

Animals and people are not normally clones. I am neither my mother's nor my father's clone, because my own genome is a mixture of genes coming from my mother and from my father. This is why I look different from my parents.

However, identical twins are clones of each other. They have exactly the same genes. This happens when the fertilized egg cell in mother's womb splits into two eggs, and each half grows into a separate and complete human. Because identical twins have identical genes, they look exactly alike. Now genticists are able to make animal clones on purpose. Remember the sheep that gives milk containing a human blood-clotting protein? Wouldn't it be quite useful to have more sheep of just the same kind—to have lots of twin sheep?

Animal clones have been made for many years. Skilled veterinarians take fertilized egg cells that have divided until there are a few dozen cells out of the mother animal. The veterinarians split them into single cells to let each develop into a clone (or identical twin) and put them back into foster mothers. Veterinarians can also store the cells at very low temperatures and save them for years.

However, genetic engineers have discovered a new trick to make animal clones. In 1997 Scottish geneticists, for the first time, showed that a clone can, in principle, be made from an adult animal, probably even from a human. They removed the cell nucleus (with all its DNA) from a sheep's egg cell. Into the empty egg cell they put a nucleus taken from the udder of another sheep. This engineered egg cell was placed into a mother sheep. It developed into a healthy clone of the sheep that had given the udder cell. The geneticists called this sheep Dolly. Dolly had exactly the same genes as the sheep that gave just one cell. Dolly was the identical twin of a sheep six years older than herself!

This was the first time ever a father was not needed to give birth to a mammal—only mothers and skillful geneticists. Most importantly, this also proved that any cell in an adult organism still has all the genes needed to make a new animal.

Of dachshunds, poodles, and *genetic* engineers

Living beings can inherit and pass on only features that are described in the genes. It would be quite useless for a mother or father to study genetics in order to make their future daughter a great scientist or to exercise often to make their future son a great athlete. We cannot change the genes we already have. We inherit and pass on features we are born with, such as the color of our skin and eyes or the size of our body.

The same applies to all other species. Breeders of animals and plants have known this for thousands of years—but, of course, they didn't know about genes and how they work. Dog breeders, for example, selected those puppies that appeared to be especially strong, fast, clever, or faithful or even had the shortest legs. Short-legged dogs were then bred with other short-legged dogs, fast dogs with other fast dogs. They passed on their features, and various dog breeds appeared. For example, dachshunds can chase rabbits out of their holes, greyhounds can run extremely fast, and poodles play very clever tricks.

Why did some puppies have shorter legs in the first place?

This happened because none of the puppies had identical genes. Even when each one had the same father and mother, each puppy ended up with different genes from their parents. The different gene combinations gave the puppies different spots on their fur and slightly different leg lengths.

When DNA is copied often, tiny mistakes called mutations may occur. For example, mutations can change the many genes responsible for making

short, long, or crooked legs. If such a mutation happens within the egg or sperm cells, it will be passed on to the offspring. The very first dachshund was a mutant. The breeder seemed to be happy with it. His short-legged dog could creep into holes and chase out rabbits. So the breeder wanted more dogs like this.

With cows, breeders made an effort to cross animals that gave more milk or behaved more patiently than others while being milked. Apple breeders selected and grew kinds of apple trees that produced bigger and sweeter apples. Our modern kinds of vegetables were also designed by breeders. They selected carrots that tasted better and grew taller than others, and decided to grow more of them. By selecting and crossbreeding, farmers changed the genes of many kinds of fruits and vegetables that we can buy and eat today.

Genetic engineers, however, do not wait for a mutation that might or might not occur. They try to find out what genes are responsible for a particular feature. In the laboratory, geneticists isolate those genes and try to build them into the DNA thread of an animal or plant. In some ways, modern genetic engineering is just a shortcut for something people have been doing ever since they began to grow plants and breed animals—selecting and propagating favorable genes.

Genetic *detectives*

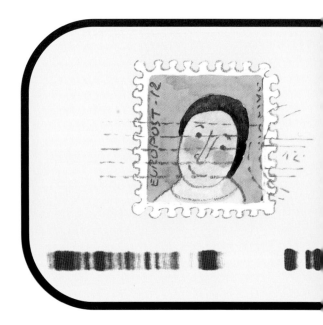

Our genes make us a member of the human family and, at the same time, make us a unique person. Most of our genes are indistinguishable from those of our brothers, sisters, neighbors, and even humans living on other continents. Our genes are even very similar to those of the chimpanzees, our closest relatives in the animal kingdom.

It took geneticists quite a long time to find the tiny differences in our large genome that make each of us different. They have learned to use these tiny differences to identify people.

At some rare spots on our DNA, each person has his or her own DNA sequence pattern—just as each of us has his or her own fingerprints. Genetic engineers can make genetic fingerprints visible and help the police to identify both victims and villains beyond a doubt. From tiny amounts of DNA left by someone, genetic detectives can find out whether a particular person has been at a particular place. These pieces of DNA can be found at the root of one's hair, in bits of skin, or in a drop of blood or saliva.

Take, for instance, the case of Mr. XY, who is suspected of sending a blackmail letter. He denies having committed the crime. However, the blackmailer has left his genetic fingerprint on the stamp of the blackmail letter. From the dried saliva, detectives can isolate DNA. If there is too little DNA to make a genetic fingerprint, a PCR (polymerase chain reaction) machine is used to make more, identical copies of the DNA.

Using enzyme scissors, genetic detectives cut and sort pieces of the DNA according to their length. Then they make visible the DNA regions that vary between individuals. Now patterns appear on an X-ray film that can be compared with the patterns obtained from all

suspects. If the blackmailer's pattern is identical to the pattern obtained from Mr. XY's saliva but distinct from any other samples, Mr. XY—with extremely high certainty—is the blackmailer.

What does a genetic fingerprint look like, Gene?

Here, I can show you some photos. This one shows the genetic fingerprint of the person that licked the stamp on the blackmailing letter. The other one shows the fingerprint of Mr. XY and of two other suspects. Now, can you tell who licked the stamp?

Genes are natural

We've reached the end of our expedition. It is time to land and enlarge ourselves to see what happend to the cut in the finger. All the genes and proteins involved in healing the wound have done their work, and the wound is closing. In this case, there is no need for genetic engineers to help.

However, millions of people suffer from hemophilia and hundreds of other diseases that occur when particular proteins are missing or damaged. Geneticists and doctors are busy searching for the genes that, when faulty, would cause such diseases. They hope to engineer human genes into bacteria and animal cells. First, they want to study the effects of the particular genes. Then, they hope to find better treatments.

Is this all natural, Gene?

Well, this is a tricky question. Many people are worried about what genetic engineers are doing, thinking it is not nature's way. However, I think it is only natural to help people in need. Hemophiliacs, for example, have the right to receive proper medicine. If this medicine can be produced by bacteria cells or in sheep's milk, why not do it?

Also, don't forget that genetic engineering was not invented by genetic engineers. It was discovered by them. In nature, we see many examples of what we can eventually achieve by using genetic engineering. Nature invented clones—identical twins. Nature created viruses that can transfer genes into our cells. Nature created the plasmids that are so tremendously useful for multiplying DNA.

Since we can alter the genes of rice plants so that rice contains more proteins and can provide people with a healthier diet—what's wrong with that? Even before anybody knew anything about genetics, farmers created plants and animals with genetic combinations that had not been seen on this planet before. Just remember the many kinds of apples you know of—and the hundreds of dog breeds.

There might be a chance to breed rice with more nutritious proteins in the traditional way. However, farmers would have to experiment perhaps for decades until by chance a rice plant with the right genes appeared. Genetic engineers, however, can do this on purpose— and much faster.

Could it be dangerous to eat plants with new genes?

No. Any apple or steak you digest consists of cells, and cells have genes. All the DNA contained in food will be digested into tiny pieces. So foreign DNA cannot affect our own DNA, which is hidden deep inside the nucleus of our cells. It just doesn't matter if the genes we eat come from a wild berry, or if they have been changed by farmers through crossbreeding or by genetic engineers.

So why are people afraid of genetic engineering?

Most people just know too little about genes. They know only that genes shape our life in some way. They are afraid that recombining genes might produce monsters or terrible bacterial weapons. Ever since mankind discovered new tools and techniques, there has been a risk of misuse. However, nature has not only given us the ability to understand genetics and all the tools to use this powerful knowledge. It has also given us the intelligence to use this powerful knowledge responsibly. It is perfectly all right to talk about what geneticists should do—and what they should not do. The more we know about genes, the better we can talk about such questions. Now we know quite a lot about genes and genetic engineering, don't we?

What do you remember?

Genes are contained in

a) all the cells of our body
b) bacteria
c) proteins
d) apples

a), b), and d) are correct. All living beings and their parts consist of cells. Proteins do not contain genes, but they are determined by the genes.

Genes

a) are recipes for making proteins
b) are stored in the cell nucleus in DNA
c) consist of chemicals called nucleotides
d) are something we eat every day

a), b), c), and d) are correct. Genes are in the cells of all living beings, telling the cells what to do. We also eat the genes of other beings that are, for example, contained in apple cells. However, we completely digest foreign genes. Our own genes are freshly built within our cells.

The genome is the name for

a) the DNA ladder that contains the genes
b) the tiny protein factory within the cell
c) all genes of a particular being

c) is correct. For example, all humans genes are called the human genome. The shape of the DNA ladder is called the double helix. The protein factories within a cell are called the ribosomes.

An identical twin is a clone

a) of his or her twin brother or twin sister
b) of his or her father or mother

a) is correct. Identical twins have identical genes, because they both developed from only one fertilized egg cell. A copy of the genome (all genes) that were contained in this first cell are in all the cells of both twins.

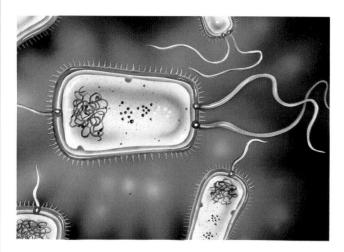

Bacteria have

a) the same genes as people
b) the same genetic code as people

b) is correct. The genetic code is the way that DNA letters (nucleotides) translate into protein letters (amino acids). A group of three nucleotides form a DNA word that stands for a particular kind of amino acid. A whole sentence of DNA words (a gene) stands for a particular sequence of amino acids that makes a particular protein. This genetic code works within all cells and within all living beings. Different living beings have different genes. Their cells produce different kinds of proteins.

Enzymes are special kinds of

a) proteins
b) genes
c) sugars

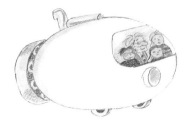

a) is correct. Enzymes are proteins that allow the chemicals within our cells and body to combine or to separate in very special ways. Restriction enzymes, for example, can cut the DNA at particular sites. Digestive enzymes break down the food we eat into smaller pieces that our cells can use.

Our liver cells and skin cells have

a) different genes
b) almost the same genes
c) identical genes

c) is correct. This is because all of our cells are offspring of one single cell that divided over and over again. Every time a cell divides, all the genes (called the genome) are copied, and each daughter cells receives one copy. However, in different kinds of cells, different genes are switched on. Therefore, liver cells make other proteins than do skin cells.

A faulty gene can make us ill because a cell with a faulty gene

a) cannot make the proper protein
b) may not stop dividing

a) and b) are correct. Genes are recipes for making proteins. If recipes for essential proteins are missing or wrong, the body doesn't function properly. Such proteins might be the ones that make blood clot or that tell the cells when to stop dividing. If cell division gets out of control, a cell might grow into a tumor.

Human proteins that an ill person may need can be made

a) by human cells only
b) also by cells of sheep or other animals
c) also by plant cells
d) also by bacteria

b), c), and d) are correct. All cells on our planet, including bacteria that consist of only one single cell, work more or less in the same way. They produce proteins according to their genes. All that a sheep, tobacco plant, or bacterium needs in order to make human proteins is the right human gene.

Transgenic animals or plants

a) carry foreign genes
b) pass on their genes to their offspring

a) and b) are correct.

Genetic fingerprints can be taken from a person's

a) saliva
b) fingernails
c) blood

a) and c) are correct. In order to take a person's fingerprint, genetic engineers need to have DNA contained in the person's cells. Fingernails and hair are made of a protein called keratin. Only when detectives find a hair with some hair root cells still attached can they isolate the DNA.

43

Glossary

Amino acids the building blocks of proteins. There are 20 kinds of amino acids.

Antibodies proteins that help kill germs.

Bacteria organisms that consist of one cell only. Many kinds live harmlessly on our skin and intestine. Other kinds can cause diseases when they reach the inside of the body to multiply there.

Blood carries nutrients and oxygen and many other substances to all parts of the body and removes the poisons and waste products. It also contains many different kinds of cells.

Blood clotting or coagulation makes blood become solid in case of injury to blood vessels.

Chromosomes packages of DNA containing the genes. In the nucleus of almost every one of our cells are 46 chromosomes in 23 pairs.

Clones living beings that have exactly the same genes as other beings. Identical twins are natural clones. Geneticists can also take the genes from adult animals and breed clones.

Crossbreeding animals and plants a traditional way of changing the genes of an organism in order to have, for example, sweeter apples or cows that give more milk. Breeders select and propagate favorable genes.

DNA short for deoxyribonucleic acid. Genes are lined up on these incredibly thin strings, shaped like twisted ladders, that are found in each cell.

Enzymes proteins that help combine and break down chemical substances.

Fibrin a protein that makes long, sticky fibers that form a meshwork and stop blood from flowing out of a wound.

Fibrinogen or fibrin maker a protein that turns into fibrin during an injury.

Gene medicine replaces faulty genes with proper ones.

Genes sections on the DNA. Stored in the cell nucleus, they are the recipes for making proteins. Each gene consists of several thousand code words. Each of a person's cells contains the same unique and complete set of genes.

Genetic code the way in which the chemical language of the genes is translated into the chemical language of the proteins. Three nucleotide letters stand for one kind of amino acid. Because the genetic code is universal to all living beings, bacteria and plants can produce human proteins once they are given the right human gene.

Genetic fingerprint the unique sequence on a person's DNA that is different from every other person's.

Genome the name for the set of genes of a species. The human genome consists of 100,000 genes.

Hemophiliacs those people missing the right proteins that would help stop blood from flowing in case of an injury.

Hereditary diseases illnesses that are the result of faulty genes inherited from the parents. Hemophilia is one example.

Human egg cell contains only 23 single chromosomes. In order to divide and form all the different cells an organism needs, a female egg cell needs to be fertilized by a male sperm cell bringing another set of 23 chromosomes.

Human sperm cell contains only 23 single chromosomes. Male sperm cells can fuse with a female egg cell in order to fertilize it.

Insulin a protein made by human pancreas cells that tell the liver cells when to take sugar out of the blood. Patients who suffer from diabetes lack insulin.

Messenger RNA takes a copy of the gene (of the protein recipe) to the ribosomes where proteins are made.

Mutations changes in the genes caused by spelling errors. Most mutated genes are faulty, but some are interesting and helpful. Mutated genes can be passed along to future generations.

Nucleotides molecules that build the rungs of the twisted DNA ladder. There are four kinds of nucleotides, nicknamed A,T,C, and G.

Nucleus the ball in the middle of the cell in which the genes are stored.

Plasmids small DNA circles on which bacteria store parts of their DNA.

Protein medicine provides patients with proteins that are taken from healthy persons or made by genetically changed bacteria.

Proteins the building blocks of cells and the tools that the cells use for their many kinds of activities. Proteins make cells, and many cells build an organism. There are 100,000 different kinds of proteins working together in the human body. Fibrinogen and fibrin are just two kinds.

Red blood cells carry oxygen to all parts of the body.

Restriction enzymes proteins that cut the DNA at specific sites. Genetic engineers can use them as gene scissors.

Ribosomes tiny balls within a cell that make fresh proteins according to the genes.

Thrombocytes cells that float with the blood and help to make it solid near a wound.

Transferring genes what genetic engineers do when they add foreign genes to a bacterium, a plant, or an animal. In nature, genes can get transferred by chance.

Transgenic organisms carry genes that originally come from other species. This way transgenic sheep, for example, can make human blood-clotting proteins once the right human gene has been given to them.

Tumors can arise when faulty genes tell cells to divide much faster than other cells.

Viruses tiny particles that have their own genes. Some kinds can cause diseases such as flu.

White blood cells come in many different kinds. They are constantly on the outlook for harmful germs that they destroy.

Index

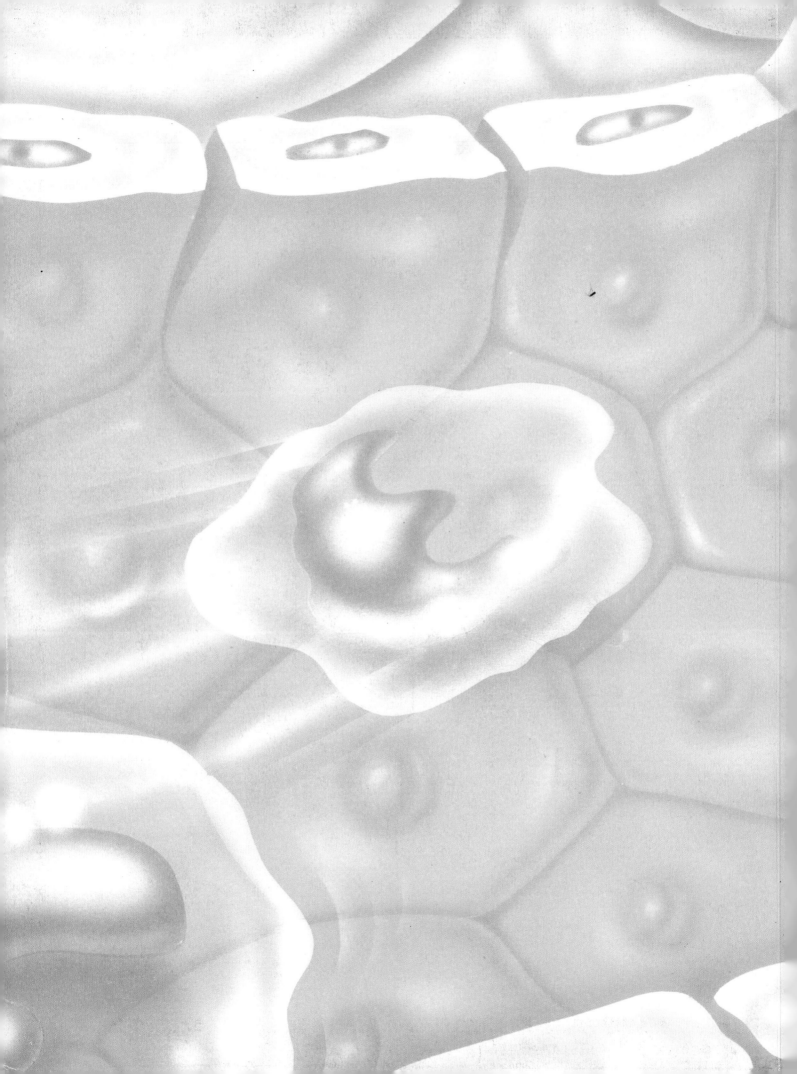